A McGRAW-HILL NEW BIOLOGY

Scientific Adviser: Dr. Gwynne Vevers
Curator of the Aquarium and Invertebrates,
The Zoological Society of London

OCTOPUS, CUTTLEFISH & SQUID

OTHER BOOKS IN THIS SERIES

Chimpanzees

Prue Napier, Illustrated by Douglas Bowness

Fishes

Gwynne Vevers, Illustrated by Alan Jessett

Frogs, Toads & Newts

F. D. Ommanney, Illustrated by Deborah Fulford

Spiders & Scorpions

J. L. Cloudsley-Thompson, Illustrated by Joyce Bee

Bees & Wasps

J. L. Cloudsley-Thompson, Illustrated by Joyce Bee

Crocodiles & Alligators

J. L. Cloudsley-Thompson, Illustrated by Joyce Bee

Birds of Prey that hunt by day

Clive Catchpole, Illustrated by David Nockels

Owls

Clive Catchpole, Illustrated by David Nockels

A McGRAW-HILL NEW BIOLOGY

Gwynne Vevers

Octopus, Cuttlefish & Squid

Illustrated by Joyce Bee

McGRAW-HILL BOOK COMPANY
New York St. Louis San Francisco

Metric Conversion Table

1 centimeter = 0·39 inch
1 meter = 3·27 feet
1 kilometer = 0·62 mile

1 sq. centimeter = 0·15 sq. inch
1 sq. meter = 10·76 sq. feet
1 hectare = 2·47 acres
1 sq. kilometer = 0·39 sq. mile

1 kilogram = 2·21 lb (avoirdupois)
1 tonne = 0·98 (long) ton

Library of Congress Cataloging in Publication Data

Vevers, Henry Gwynne, date.
Octopus, cuttlefish & squid.

(McGraw-Hill new biology)
Includes index.
SUMMARY: Discusses three members of the large group of invertebrate animals known as mollusks.
1. Cephalopoda—Juvenile literature. [1. Mollusks]
I. Bee, Joyce. II. Title.
QL430.2.V52 594'.56 77–25083
ISBN 0–07–067405–1

First distribution in the United States of America
by McGraw-Hill Book Company, 1978.

First printed in Great Britain for
The Bodley Head
by William Clowes & Sons Ltd., Beccles
First published 1977

Contents

Different types of cephalopods. They are drawn approximately one third of their natural size.

1

Introduction

Fishes, amphibians, reptiles, birds and mammals all have a backbone as well as many other bones inside their bodies. They are known as the "vertebrates," but there are many kinds of animals without backbones or indeed any bones. These are the "invertebrates." Some invertebrates, such as insects, scorpions, crabs and lobsters, have a firm outer covering which has to be shed or moulted at intervals as they grow.

Octopuses, cuttlefishes and squids belong to the large group of invertebrate animals known as molluscs or "soft-bodied animals." This group also includes snails and slugs, cockles and mussels, most of which have an outer shell enclosing the soft body. Snails have only one shell which is usually coiled, whereas cockles, mussels and clams have two shells connected by a hinge, known as "bivalves." Slugs have no shell.

Most octopuses, cuttlefishes and squids have no outer shell. Inside its body a cuttlefish has a rigid "cuttlebone," and squids have an internal horn-like "pen" which helps to support the body. (Cuttlebones are often washed up on the beach and they are used by canary breeders to supply their birds with calcium.) A few octopuses have the remnants of a shell in the form of a thin plate, but most have no trace of a shell and the only hard part of the body is the beak.

Snails have a muscular, fleshy foot on which they crawl around on the ground or on the sea floor. Cockles have a differently shaped foot which helps to anchor them in the sand. Octopuses, cuttlefishes and squids do not have this kind of foot; instead they have eight arms, and squids and cuttlefishes also possess two long tentacles. The whole appearance of an octopus gives an impression of head and arms (or feet), and so these molluscs and their relatives, the cuttlefishes and squids, are known as "cephalopods," which means "head-feet."

Cephalopods all live in the sea. Most of them move fast—some very fast. All are carnivorous, feeding on other animals, such as fishes and crabs, shrimps and prawns, which they hunt mainly by sight.

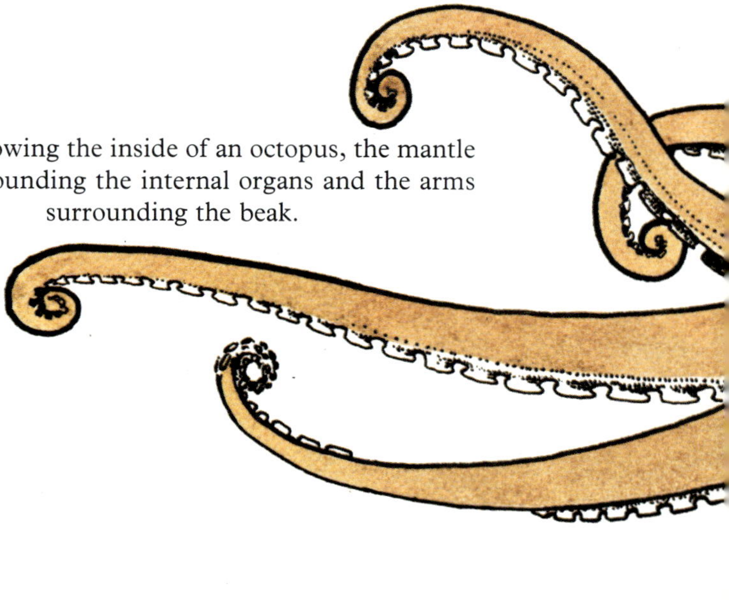

Diagram showing the inside of an octopus, the mantle muscle surrounding the internal organs and the arms surrounding the beak.

2

Octopuses

There are about 150 different kinds of octopus living in the seas of the world, mostly in tropical or subtropical areas. Some are small, only a few centimeters across, but there are a few very large ones. In the northern Pacific there is an octopus with arms that are 5 meters long, so that it has a span of about 10 meters. An octopus of this size could, in fact, hold a man under water so that he would drown.

The body of an octopus is a soft and very muscular bag, known as the mantle, containing the internal organs. Within the lower part is a water-filled space with an opening into the sea water outside, called the

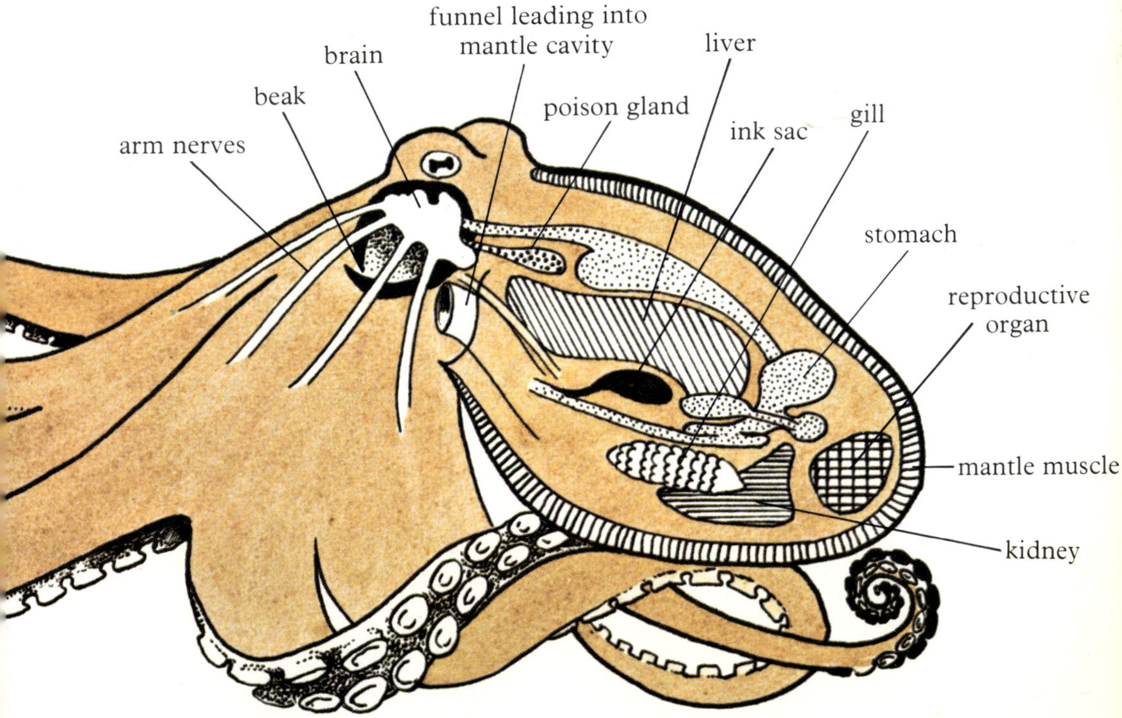

funnel. The space or "mantle cavity" contains the gills. Water enters through the funnel and flows over the gills, where it gives oxygen to the gill blood vessels and removes carbon dioxide. Most of the time the water is pumped in and out of the mantle cavity quite slowly.

The head is roundish and it has two large eyes. Underneath the head is the mouth, with a sharp beak inside. Surrounding the mouth there are eight very long flexible arms, each with a single or double row of suckers. The suckers are on stalks and can be moved in any direction. Each sucker can also be expanded to form a flat plate or contracted to form a gripping organ. With the help of these

Octopus moving slowly over rocks
on the sea floor.

suckers the arms of an octopus can pick up stones and other objects and turn them over. When the octopus is clambering about on the sea floor some of the muscular arms are extended so that they become quite thin and their suckers grip hold of a rock. Then the arms contract and so pull the body of the octopus forwards. The whole action is very smooth, without jerky movements, and quite slow.

When it wants to move fast an octopus uses jet propulsion. The muscles in the wall of the mantle cavity contract very rapidly and the water contained in the mantle cavity is forced out through the funnel in a jet so that the octopus is propelled in the opposite direction. If the opening of the funnel is held facing forwards the animal will shoot backwards. The funnel can, however, be bent so that its opening faces backwards, and the jet from the mantle cavity will then propel the octopus forwards.

It sometimes makes this rapid movement when it is within range of its prey. It can then, for instance, engulf a crab in a kind of soft flexible umbrella formed by the outstretched arms and the webs which connect them. The octopus then bites the crab with its powerful sharp beak. As it does so, some of its saliva, which contains a powerful poison, enters the wound and helps to kill the prey. The octopus then takes the prey back to a sheltered lair among the rocks where it is eaten.

Octopuses have excellent eyesight and their sense of touch is also good, and these senses together with the well-developed brain enable them to learn very quickly about the world around them. Using their eyes they can distinguish between square and circular objects and by touching with the tips of the tentacles they can tell the difference between smooth and rough objects.

Octopuses are able to change color very rapidly.

When living in shallow water among rocks and weeds they have a pattern of black, grey and white mottling, often with a greenish tinge, and they are then extremely difficult to see.

Octopus resting.

Octopus escaping from moray-eel.

When attacked an octopus usually expels a cloud of blackish ink. This is a pigment produced in its body and it is squirted out into the water through the funnel. The ink serves partly as a screen behind which the octopus can escape, and partly to distract the attacker which snaps at the ink cloud while the octopus swims away. Octopuses are also able to squeeze themselves through very narrow gaps in the rocks and this is yet another means of escaping from animals that prey on them.

In many parts of the world octopuses are preyed upon by moray-eels, some of which are about 2 meters long. These fishes hunt mainly by scent, for their eyesight is poor. Normally a moray-eel seizes one of the arms of the octopus, gripping firmly with its sharp teeth. The moray-eel then stretches its body so that it is quite straight and then twists around and around until the octopus's arm is wrenched off. After swallowing this arm the moray-eel repeats the action until the octopus is left with no arms, and its body is then eaten.

3

Octopus breeding

It is not difficult to distinguish a male octopus from the female. In the male the third right arm has an area at the end without suckers. This is known as the hectocotylus and during mating the male uses it to transfer a packet of sperms from its body into the mantle cavity of the female. This packet, known as a spermatophore, contains a group of sperms.

I have seen octopuses mating in the Aquarium of the London Zoo. We had a pair of Common octopuses which had come from Madeira. They both settled down quite well and started to feed on shore

crabs, each taking about five every day. The female spent most of her time below a rocky ledge, but the male was more active and wandered around the tank. About nine days after their arrival the male took up a position on a rock about a meter away from the female, and extended his hectocotylized arm from time to time, touching the female very gently with its tip. At first she repelled him but eventually he succeeded in placing the tip of this arm, loaded with spermatophores, inside her mantle cavity. There it remained for about an hour and during this time the spermatophores must have traveled down the hectocotylus and become attached to the female. The spermatophore would then burst and release

Mating of Common octopuses on the sea floor.

the tiny sperms so that they could move up the oviducts and fertilize the eggs. This, of course, we could not see.

The female started to lay eggs seven weeks after the mating, and continued to do so over a period of twelve days. They were laid in clusters, each with several eggs attached to a central strand. There were a number of these clusters, and this female must have laid nearly 40,000 eggs. The clusters were attached to a ledge in the tank, hanging down in what looked like very long bunches of tiny grapes. In the sea the eggs are usually laid in fairly shallow water and the clusters are attached to rocks. Each individual egg was about 3–4 mm. long, and quite translucent.

After the eggs were laid the mother remained immediately below them, clinging onto the overhanging ledge with her suckers, but always with at least one eye watching the surrounding area. During the period the mother octopus was continuously squirting water from her funnel over the eggs. In this way she ensured that the water around the eggs was always moving, never stagnant, so that the eggs received plenty of the oxygen dissolved in the surrounding water. She also ran the thin tips of her arms through the clusters, thus loosening any pieces of dirt, which would be carried away in the current.

During incubation the female draped her arms around the eggs so that they were well protected

The female octopus guarding her eggs in a tank at the London Zoo Aquarium.

against intruders. At this time the tank also contained some starfishes and sea-urchins, but when any of these came too close she pushed them away with one of her arms. She also chased away another octopus that came too close. Even more surprisingly the female also refused any food that was offered to her, and would pick up and remove it if it fell near the eggs. By these means she succeeded in keeping the eggs clean and healthy, and as a result they developed quite normally.

At first the eggs showed no sign of color, but as time went on they started to produce dark pigment, particularly around the eyes, and this could be clearly seen through the egg case.

In the early stages of development the yolk, which supplies the developing embryo with food, lay at the bottom of the egg, away from the stalk. In the later stages as the arms started to form and pigment appeared on the body, the whole embryo turned around within the egg-case, so that the yolk and the circle of arms faced upwards, towards the thin thread attaching it to the egg cluster.

The female octopus continued to protect and brood the eggs for at least forty days and then they started to hatch, over a period of about eight days. When an embryo was ready to break out of the egg case it did so at the bottom end of the egg, taking the remains of the yolk with it, which looked as though it was clasped within the circle of eight tiny, short arms.

The newly hatched young octopuses were only about 2 mm. long, and as they became free of the egg case they swam up towards the light. Then after a few hours they sank to the bottom. Many of them lived for eight days, feeding on their yolk, but although various kinds of food were offered they eventually died, and the female also died about a week after the last young had hatched. She had scarcely fed at all for well over a month and her arms and body had become very thin. It seems that the female octopus always dies after she has successfully guarded and hatched her enormous brood of eggs.

In the wild the thousands of tiny young would soon be dispersed by water currents, and most of them would very quickly be eaten by small fishes and other animals. Perhaps only a few hundred would survive for twenty-four hours, and only a few would live long enough to breed themselves, but there would be plenty to keep the population constant.

Newly hatched octopuses: the drawing in the box shows their actual size.

In the sea, mating takes place on the bottom among rocks, occasionally in quite shallow water, and not many people have seen it. It seems that the behavior of the pair may vary according to the type of octopus. For instance, in the Reef octopus of southeast Asia, the male has been seen to rise out of his lair in the form of a vertical column. The female was in her hole about 60 cm. away, with some of her

Reef octopuses mating.

arms spread out on the sea bed. The male changed his pattern to show black stripes along the arms on the side facing the female, but transverse bars on the other side. He stretched out his third right arm, the one with the hectocotylus, towards the female and inserted it into her mantle. They remained in this position for about ten minutes, keeping quite still.

4

Paper nautilus

Although most octopuses live in shallow water, usually among rocks or in coral reefs, there are some in the open sea.

One of these is *Argonauta*, known as the Paper nautilus, because the female builds a paper-thin shell around her body. She is about 30 cm. long, including the arms, but the male is only 10 mm. long. In fact it was many years before the male Argonaut was discovered.

Argonauts are not often seen, but they probably live most of the time at the surface of the sea, where

air caught within part of the shell helps to keep them afloat.

The tiny male Argonaut has no shell but in among the arms it has a small sac containing the coiled hectocotylus, which is the third left arm. When the sac bursts the hectocotylus is released and it then unwinds to a length of up to 12 cm. or many times the length of the male Argonaut.

The hectocotylus enters the mantle cavity of the female carrying with it the sperms which will later fertilize the eggs. At this stage the hectocotylus breaks from the male's body and remains within the female. The eggs are attached to the papery shell, and guarded by the female who shoots jets of water over them.

Female Paper nautilus holding her shell with the suckers of two of her arms.

5

Cuttlefishes

Cuttlefishes mostly live near the bottom of the sea in shallow coastal waters, where they hunt for prawns, shrimps and small fishes. A cuttlefish has a head with a mouth and two large eyes, and behind this is the body. Around the mouth there is a ring of eight arms, within which there are two tentacles, normally contracted and quite short.

The squat body of a cuttlefish is supported by an internal shell or cuttlebone, and below this lie the stomach and liver. Below this again is the mantle cavity containing the gills, which opens to the outside by the funnel. Sea water enters the mantle cavity through the funnel and bathes the gills. As in the true fishes the gills are delicate organs which have many blood vessels. The oxygen dissolved in the sea water passes into these blood vessels, just as

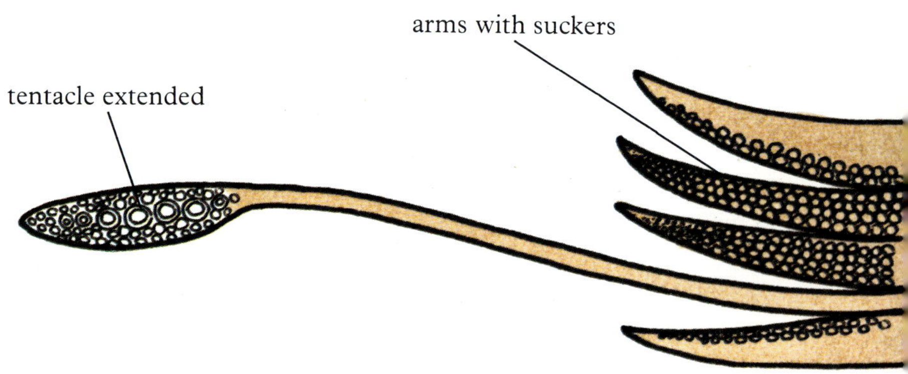

Diagram showing the inside of a cuttlefish.

the oxygen in the air passes into the blood vessels in the lungs of a bird, a cat or a man.

The mantle cavity has another important task. Its walls are very muscular and when the cavity is filled with water it can be emptied by the very rapid contraction of these muscles. When this happens the water is forced out through the funnel in a jet, and as in the octopuses this has sufficient force to propel the cuttlefish backwards.

Unlike octopuses, cuttlefishes have a long fin on each side of the body. The fins make a wavy movement which can move the animal along quite slowly. The fins can move independently of one another, and if one undulates forwards and the other backwards the animal will start to turn on its own axis. So with jet propulsion for fast movement and the fins for slow maneuvers the cuttlefish is very efficient in the water.

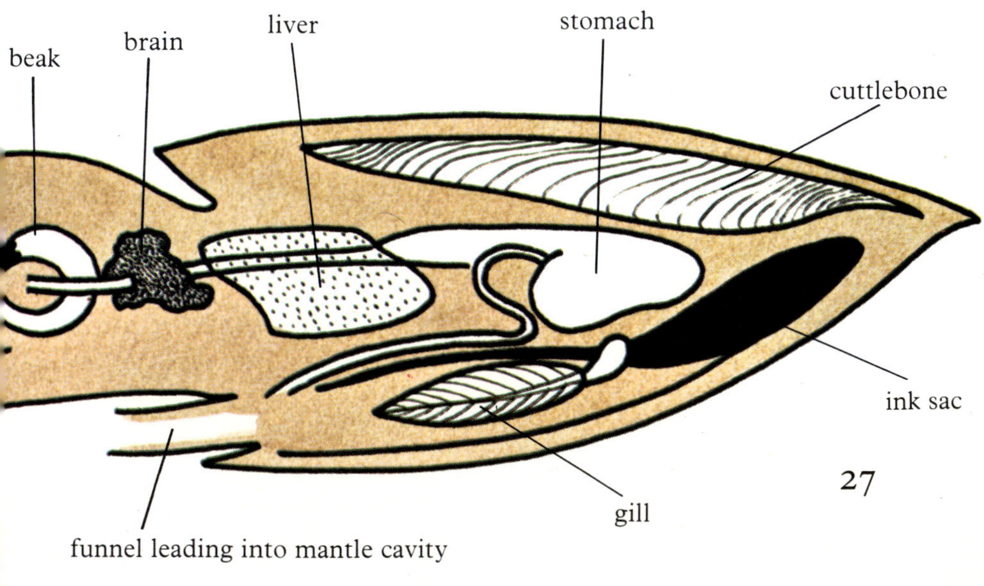

The cuttlebone lying beneath the skin of the back is a fairly soft, chalky structure which supports the soft parts of the body. Perhaps even more important, the cuttlebone helps the cuttlefish to remain afloat. When examined under a lens it can be seen to contain numerous very fine channels. The cuttlebone is light in weight when these channels are filled with gas, but becomes heavier when some of this gas is replaced by fluid. This allows the animal to remain afloat at any depth in the water or to sink and hide among weeds on the bottom.

Cuttlefish catching a prawn on the sea floor.

The large eyes have lenses as in man and other vertebrates which can also be focused onto nearby objects. The information picked up by the eyes is passed to the brain which sends messages along the nerves to other parts of the body. When the eyes see suitable prey, usually a prawn, they pass this information to the brain which sends a message to the two tentacles which are then shot out to catch the prey.

Three ways in which cuttlefish change color.

Like other cephalopods the cuttlefishes have an astonishing ability to change color very rapidly. The pigments that provide them with colors are mainly yellow, red, brown and black granules and they are held in thousands of tiny sacs in the skin. Muscles attached to their walls can expand or contract these sacs. When the sacs are expanded the pigments will be spread out and the colors become darker, but when contracted the granules of pigment will be gathered into a tiny mass, so that little color is seen and the cuttlefish will appear pale.

Cuttlefishes are very good at concealing themselves by changing their coloring to match their surroundings. On a sandy bottom a cuttle is usually quite pale, but on a rocky bottom it shows patches of dark and pale skin which help to break up the outline of the body, so that it becomes quite difficult to recognize.

At other times the pigments may produce a more striking pattern. This happens when the cuttlefish suddenly sees a large strange animal. The whole of its body then becomes very pale, except for two circular black spots on the back, looking rather like large eyes. This probably scares the intruder and makes it move away.

The ink produced by the Common cuttlefish (*Sepia officinalis*) has been used by artists since long ago to produce the rich brown pigment known as sepia.

6

Cuttlefish breeding

Cuttlefishes mate in spring and summer. They recognize each other by sight. A male that is ready to breed has a distinctive pattern of vertical zebra-like stripes of yellowish-white and purplish-brown. When another cuttlefish approaches he flattens out one of the arms on the side nearest the newcomer. This is a form of display and if the second cuttle has

the striped pattern and shows the flattened arm it too will be a male. The smaller or weaker of the two will then retreat. If, however, the second cuttlefish does not display in this way the male recognizes it as a female and remains with it. Just before mating the male turns so that he is facing the female and then he comes close to her and uses his hectocotylus (in cuttlefishes the fourth left arm) to place a spermatophore in a small pouch just below the female's mouth. Mating in this way may take place several times a day.

Cuttlefishes mating (the male is on the right).

Cuttlefish eggs on seaweed.

Quite soon the spermatophores burst, releasing large numbers of sperms which move into the mantle and oviducts where some of them fertilize the eggs. In the Common cuttlefish (*Sepia officinalis*) the female lays the eggs one at a time, each being enclosed in a black casing about the size of a small grape. The eggs are attached to a rock or to a piece of seaweed, but the female, unlike the female octopus, pays no further attention to them.

7

Squids

Squids have an elongated body supported by a thin transparent horny "pen." The mouth is surrounded by eight arms and there are also two long tentacles which can be shot out to catch prey. There are many different kinds of squids that swim freely in the open sea. Some are just a few centimeters long and these mostly live in the surface waters. The Common squid of the Mediterranean, Atlantic and North Sea has a body about 20 cm. long with eight arms and two tentacles. There are others with a body up to 4 meters long and with arms extending to 15 meters or more. These are the giant squids.

As in the other cephalopods the internal organs, such as heart, guts, and reproductive organs, lie above the mantle cavity. Squids move by rapidly contracting the mantle cavity so that the water in it is expelled in a jet, in much the same way as the octopuses and cuttlefishes. The jet leaves the mantle

Squid "pen."

cavity through the funnel. In this way squids can move extremely rapidly both forwards and backwards.

Squids also have a fin on each side of the mantle and this is used for slow movements in either direction. When swimming in open water, Common squids are almost transparent, sometimes with a pink tinge on the upper surface. They can change color to produce different patterns, but not so strikingly as the cuttlefishes.

Some of the smaller squids can apparently remain floating without actively swimming. This means that some part of their body must be lighter than the sea around them. In fact, quite recently it has been found that certain parts of the body of these squids are filled with a solution of ammonium chloride. This is a liquid that is not as heavy as sea water, and gives the squids buoyancy. The ammonia is a waste product which in other squids is excreted into the water. One of these squids was seen to float with the head down, so that the arms were hanging like a curtain in the water, and perhaps this helps them to catch passing prey.

Squids are eaten in large numbers by sharks and other fishes, by many toothed whales including sperm whales and pilot whales, and also by some sea birds, such as penguins. The stomach of one Bottle-nosed whale was found to contain about 10,000 squid beaks.

8

Squid breeding

Male squids in breeding condition have prominent white flecks on the sides of the body and a white band running along the underside of the mantle. When another mature male approaches, this pattern becomes more distinct and it seems to act as a signal to keep the males apart, for the smaller of the two males usually moves away. On the other hand, when a female approaches a male she does not make this signal, and the two squids will then remain together, sometimes for days. Eventually the male succeeds in leaving spermatophores within the mantle cavity of the female.

The female lays her eggs in a string, embedded in a kind of jelly, and these are usually attached to rocks on the sea bottom. Often several females lay their eggs together in the same area, but each string is attached separately. These eggs are not protected in any way by either the male or female squid.

Strings of squid eggs attached to rocks on the sea floor.

In the shallow water off the western coast of North America enormous numbers of the Pacific coast squid *Loligo opalescens* gather for mating. They mostly come together in pairs, but sometimes in groups of four or five, and during this time they are much preyed upon by sharks and other large fishes. During this very active period of courtship their tentacles intertwine and their colors change continuously, but are mainly translucent red and gold. During the actual mating the squids sink to the bottom, the male seizes the female from below and

Pacific coast squids gather for mating.

inserts his hectocotylus, which holds the spermatophores, into her mantle cavity. The eggs are later laid in jelly-like capsules, over a period of three or four days. Each capsule, containing 70–100 eggs, has a thin filament at one end. This is used to attach the capsule to some object on the bottom, and this prevents it being moved about by the water currents.

The adults die after mating and egg-laying, leaving the eggs unprotected on the sea floor. When they hatch the young are about 3 mm. long.

9

Squids in deep water

The smaller squids are usually caught in trawl nets, sometimes in large numbers, and mostly in fairly shallow water. The Japanese, for instance, catch over half a million tons of the squid *Todarodes pacificus* every year.

In deeper water there are several other squids, some very large. Perhaps the best known is the giant squid *Architeuthis*, which probably lives at depths down to 3,000 meters. In 1878 a dead specimen was washed up in Newfoundland. It had a body at least 5 meters long and the two tentacles were about 12 meters long. This is probably one of the squids that are caught and eaten by sperm whales, which are up to 20 meters long. The squids must put up a good fight, for the scars left by their suckers are often found on the body of the whale, particularly around the mouth. It seems that such deep-water squids may occasionally be swallowed alive, for sucker marks which could only have been made by a live squid have been found on the inner lining of a whale's stomach.

In fact, some of the squids from deep water have never been seen alive and are only known from their remains found in the stomachs of whales or of large fishes. One of these, taken from the body of a dead

A deep-sea squid which produces light, *Lycoteuthis diadema* (twice natural size).

sperm whale, was *Lepidoteuthis grimaldii*, which had a body covered with scales, rather like a reptile. The attractive squid *Entomopsis alicei* has been found on several occasions in the stomach of fishes caught in the Atlantic Ocean, west of Portugal.

These large squids are probably capable of moving at great speed, and this may be one of the reasons why they have never been caught.

Many of the squids are luminescent, that is, they are able to produce light. Some of them have small organs, known as photophores, in which the light is produced. For instance the deep-sea squid *Lycoteuthis diadema* has five of these photophores around the rim of each eye, two on each tentacle and eight on the body. Some of these photophores give a blue light often with a pearly sheen, while others produce a reddish tinge. Perhaps the light from the photophores helps these squids to recognize one another, particularly during the breeding season, but it is also possible that they act as searchlights to light up the surrounding water, and perhaps even to attract prey.

The squid *Chiroteuthis lacertosa*, from depths of about 1,000 meters in the North Atlantic, has two very long tentacles, each with photophores and a group of hooked suckers at the tip. It has been suggested that these tentacles may be lowered below the body, so that the photophores can act as lures to attract prey which is then caught by the suckers.

10

Fossil cephalopods

Although most of the living cephalopods have no shell, there was a time many millions of years ago when their ancestors did live in shells. This is known because the shells, but not the soft parts of the body, have been preserved in the rocks as fossils, in the same way as the bones of dinosaurs and many other animals that are now extinct.

Chiroteuthis lacertosa with its long tentacles that produce light (natural size).

Fossil nautiloid, with half the stone cut away to show the "chambers" or sections made as the animal grew. (Courtesy the Trustees of the British Museum, Natural History)

The best known of these fossil cephalopods are the nautiloids and ammonites which flourished about 170 million years ago. They had a coiled shell, which presumably was quite light. In the fossils the shell has been changed into stone and so a fossil ammonite is usually quite heavy to hold.

The ammonites and most of the nautiloids died out a long time ago but one of the nautiloids, known as the Pearly nautilus, has survived.

11

Pearly nautilus

Nautilus, which is now found only in the tropical parts of the Indian and Pacific Oceans, has a rather complicated coiled shell with a series of chambers. As it grows these chambers increase in number and size and the animal lives in the last one, which is the largest. The other chambers contain gas which helps to keep the animal afloat.

Pearly nautilus, with one of its eyes showing. It has forty to fifty arms with long, thin ends.

Nautilus has forty to fifty arms, but these do not have suckers. It moves by jet propulsion, squirting water out through the funnel, but it is a clumsy, slow-moving cephalopod compared with the streamlined squids. In fact, over a period of many millions of years the cephalopods have become increasingly active compared with their slow-moving relatives the snails, slugs and bivalves. Most of them have, for instance, lost the external shell which must have been an encumbrance for an active animal. Most of the octopuses and cuttlefishes have remained in shallow waters, where by changing color they can hide from enemies. The squids, on the other hand, have colonized the open sea, often down to great depths. They have developed some astonishing techniques for remaining buoyant and for producing light, and above all they have become some of the fastest and most efficient predators in the sea, well able to compete with fishes of comparable size.

Index